The Valuation of Wetlands

AN APPRAISAL INSTITUTE HANDBOOK

by

David Michael Keating, MAI

**APPRAISAL
INSTITUTE®**

875 North Michigan Avenue
Chicago, Illinois 60611-1980

Vice President, Communications: Christopher Bettin
Manager, Book Development: Michael Milgrim, PhD
Editor: Janet Seefeldt
Graphic Designer: Claire Baldwin

For Educational Purposes

The material presented in this text has been reviewed by members of the Appraisal Institute, but the opinions and procedures set forth by the author are not necessarily endorsed as the only methodology consistent with proper appraisal practice. While a great deal of care has been taken to provide accurate and current information, neither the Appraisal Institute nor its editors and staff assume responsibility for the accuracy of the data contained herein. Further, the general principles and conclusions presented in this text are subject to local, state and federal laws and regulations, court cases and any revisions of the same. This publication is sold for educational purposes with the understanding that the publisher is not engaged in rendering legal, accounting or any other professional service.

Nondiscrimination Policy

The Appraisal Institute advocates equal opportunity and nondiscrimination in the appraisal profession and conducts its activities without regard to race, color, sex, religion, national origin, or handicap status.

Library of Congress Cataloging-in-Publication Data
Keating, David Michael.
    The valuation of wetlands / by David Michael Keating.
        p. cm.
    Includes bibliographical references.
    ISBN: 0-922154-21-X (pbk.)
    1. Wetlands—Valuation—United States. 2. Wetland conservation—Economic aspects—United States. 3. Real estate development—Environmental aspects—United States.
    HD1671.U5K43   1995   95-30159
    333.91'8—dc20   CIP

Table of Contents

Readers of this text may be interested in the following related books from the Appraisal Institute: *The Appraisal of Real Estate,* tenth edition; *Environmental Site Assessments and Their Impact on Property Value: The Appraiser's Role; Real Estate Valuation in Litigation,* second edition; and *Measuring the Effects of Hazardous Materials Contamination on Real Estate Values: Techniques and Applications;* and the videotape, *Hidden Factors: Environmental Risk Evaluation and the Real Estate Appraiser.*

For a catalog of Appraisal Institute publications, contact the P/R Marketing Department of the Appraisal Institute, 875 N. Michigan Avenue, Ste. 2400, Chicago, IL 60611-1980.

Foreword

Numerous property types as well as specific valuation tools and techniques lend themselves to incisive study in short, concise publications. With the publication of *The Valuation of Wetlands*, the Appraisal Institute initiates a series of handbooks designed to address such topics.

While wetlands were once considered of negligible if any value, in recent years they have attracted the attention of the environmental movement and are now viewed as a valuable ecological resource. Appraisers can find themselves caught between opposing viewpoints and methodologies for estimating the value of a wetland. This handbook will help appraisers and other real estate professionals sort out these issues as well as provide them with a basic understanding of the physical characteristics of wetlands.

The Appraisal Institute welcomes reader suggestions for other topics to be covered in its handbook series. Comments may be directed to the attention of the senior vice president of Communications at the Appraisal Institute's national headquarters in Chicago.

As the leading publisher of appraisal literature, the Appraisal Institute is pleased to initiate this handbook series with *The Valuation of Wetlands*, and we look forward to publishing additional titles under the handbook series banner.

Richard C. Sorenson, MAI
1995 President
Appraisal Institute

Acknowledgments

I acknowledge the following for their contribution to this monograph: Christopher Bettin for seeing promise in this project and guiding me through it; Rhodes Robinson of Environmental Services, Inc., for the series of photographs on wetland creation; the anonymous reviewers on the Publications Committee for their helpful comments; Jan Seefeldt for her editorial assistance; and Walter M. Lampe, MAI. Also I acknowledge the support of my wife and recognize the sacrifices she endured as I worked nights and weekends on this manuscript; my parents for their constant encouragement; and the Lord for His grace.

David Michael Keating, MAI
April 1995

About The Author

David Michael Keating, MAI, is a graduate of the real estate program at the University of Florida. He has over eight years of real estate appraisal and consulting experience and specializes in analyzing unique properties. He has published various papers on real estate issues. He was appointed to the Appraisal Institute's Young Advisory Council in 1994 and 1995. He currently resides in Jacksonville, Florida, with his wife Amy and young son Christopher.

Introduction

Recent scientific discoveries, a changing legal-political landscape, and a growing environmental movement have radically altered the status of wetlands in the marketplace. As a result, market participants are struggling to adapt to these changes, and the valuation of wetlands has become a controversial and often confusing process.

The current attention to wetlands, their elevation to an important natural resource, and the controversy concerning their valuation are relatively recent market developments. Prior to the 1970s wetlands were viewed as little more than a nuisance, and their drainage and conversion were even encouraged by the federal government through Swamp Acts, low interest loans, and subsidies in an attempt to stimulate agricultural productivity and frontier development. As a result, 117 million acres of the nation's original inventory of 221 million acres of wetlands have been lost[1] (see Figure 1). According to the U.S. Fish and Wildlife Service, "based on current estimates, the Nation will lose an additional 4.25 million acres of wetlands by the year 2000."[2] If this statistic proves true, less than 100 million acres of wetlands, or 45% of the original inventory, will exist at the beginning of the twenty-first century.

The greatest percentage of wetland depletion has occurred in those states located along the eastern seaboard, in the South, and adjacent to the Great Lakes and in California. However, every state is home to wetlands and has been affected. At least 22 states have lost more than half of their wetland inventory[3] (see Figure 2).

1. *A National Program for Wetlands Restoration and Creation.* Report of the Interagency Committee on Wetlands Restoration and Creation (August 1992).

2. *Wetlands: Meeting the President's Challenge* (Washington, D.C.: U.S. Department of the Interior, Fish and Wildlife Service, 1990).

3. T. E. Dahl and C. E. Johnson, *Status and Trends of Wetlands in the Coterminous United States, Mid-1970s to Mid-1980s* (Washington, D.C.: U.S. Department of the Interior, Fish and Wildlife Service, 1991).

FIGURE I
U.S. Inventory of Wetlands 1780-1980

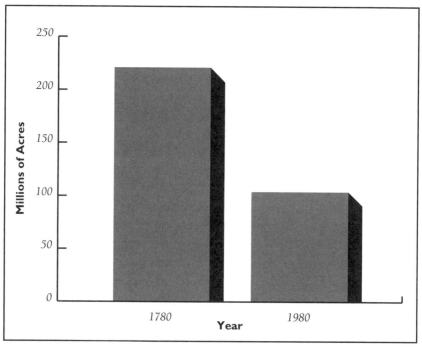

Recent scientific discoveries have changed society's perception of wetlands and their value in the marketplace. Wetlands are now known to provide critical water emergence, water reclamation, flood buffer, erosion resistance, wildlife habitat, recreation, and aesthetic functions. As a result, the U.S. government has reversed its earlier policies and is now attempting to protect and conserve wetlands. This radical reversal culminated in the Executive Order signed by President George Bush on August 9, 1991, establishing a goal of "no net loss" of wetlands. Wetlands have been transformed from an undesirable land form to an important natural resource, and they are currently one of the most highly regulated types of real estate. This change in status has had a powerful impact on the development potential and resultant market value of wetlands.

Wetland valuation requires consideration of many disciplines including biology, chemistry, social sciences, economics, and state and federal law.[4] At a

4. Kathryn Cowdery, Karl Scheuerman, and J. Christopher Lombardo, "The Valuation of Wetlands," *Journal of Land Use and Environmental Law*, vol. 1, no. 1 (Florida State University), p. 1.

Introduction

2

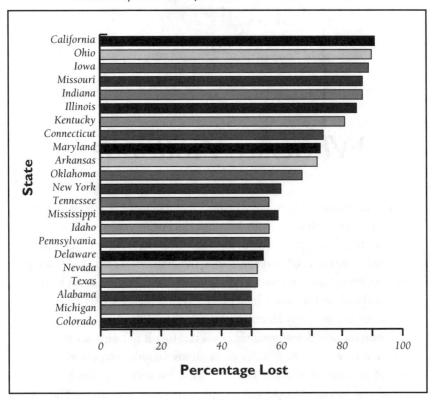

FIGURE 2
Wetland Depletion Rates for States with 50% or more losses

minimum, competent valuation involves a basic understanding of the physical, legal, economic, and social characteristics of wetlands. This handbook was written to provide appraisers and other market participants with a basic understanding of the current status of wetlands and to meet a need in the market for better information concerning the wetland valuation process.

The handbook is divided into five chapters, excluding this brief introduction and background information. Chapter 1 introduces the federal definition of wetlands as well as the five basic types of wetland systems. Chapter 2 discusses the physical, legal, social, and economic characteristics of wetlands. Chapter 3, Controversial Values, explains some of the reasons for the widely diverging values being assigned to wetlands. Application of the sales comparison, cost, and income approaches to wetland valuation is discussed in Chapter 4, and the final chapter, Future Trends: Wetlands Mitigation Banking, presents a new development in the wetland marketplace.

Chapter 1

What Are Wetlands?

The term *wetland* brings to mind distinct images of swamps, bogs, marshes, and other areas with wet mucky bottoms, thick with grass, and filled with fish, wildlife, and insects. Indeed, most wetlands share these characteristics. However, some are disguised and during the dry season aren't wet at all. But when the rains come and the snows melt, they become inundated with water and take on the classic look of a wetland. Thus, since wetlands are not always wet, properly identifying them can be difficult.

Identifying and delineating wetlands is difficult not only for the layman; even experts have differing opinions. In an attempt to clarify some of the confusion over "what is" and "what isn't" a wetland, the U.S. government has established specific guidelines and definitions. The 1986 Emergency Wetlands Resource Act defines a wetland as

> land that has a predominance of *hydric soils* and that is *inundated or saturated by surface or groundwater* at a frequency and duration sufficient to support, and that under normal conditions does support, a prevalence of *hydrophytic vegetation* typically adapted for life in saturated soil conditions.[1] (emphasis added)

This definition specifies three primary characteristics of wetlands: 1) hydric soils, 2) wetland hydrology, and 3) hydrophytic vegetation.

HYDRIC SOILS

The *Federal Manual for Identifying and Delineating Jurisdictional Wetlands* defines hydric soils as those that are "saturated, flooded, or ponded long

---

1. Emergency Wetlands Resources Act of 1986, 16 USC 3901.

enough during the growing season to develop anaerobic conditions in the upper part."[2] Anaerobic conditions result from a lack of oxygen. Therefore, hydric soils are soils that lack oxygen in the upper layers because they are covered with and retain water at some point during the year. Generally, such soils have a low rate of percolation and consist of material such as clay and muck.

For example, in the summer an area may experience more rainfall than in other seasons. Rivers may rise and eventually spill over existing banks and flow into adjoining low-lying areas. If the spillover water remains on the low-lying areas long enough to create anaerobic conditions, then the area is most likely a wetland. A soil scientist is best qualified to make such determinations based on soil surveys, core borings, and other such information.

HYDROLOGY

Hydrology is the "science dealing with the properties, distribution and circulation of water on the surface of the land, in the soil and underlying rocks, and in the atmosphere."[3] Thus, hydrology is the study of the movement of water. Wetlands are characterized by a specific hydrology, for they are periodically saturated and inundated with water, which in turn deprives the soil of oxygen and creates anaerobic conditions. Hydrologists study flood maps, geographic formations, changes in elevation, aerial photographs, the depth and flow of underlying water tables, and similar data to determine if a property is a wetland.

HYDROPHYTIC VEGETATION

Lastly, wetlands are characterized by vegetation that can grow despite anaerobic soil conditions and periodic water inundation. Such vegetation is termed *hydrophytic,* and some species may grow in the soil while others may float in the water. Common examples of hydrophytic vegetation include cypress trees, cattails, and water lilies. The federal government has established a list that can be used in identifying and delineating wetland vegetation. A trained biologist with a working knowledge of the plant species on this list is best qualified to make such determinations.

The actual identification and delineation of wetlands should be left to

2. Federal Interagency Committee for Wetland Delineation, *Federal Manual for Identifying and Delineating Jurisdictional Wetlands.* (Washington, D.C.: U.S. Army Corps of Engineers, U.S. Environmental Protection Agency, U.S. Fish and Wildlife Service, and U.S.D.A. Soil Conservation Service, 1989), p. 6.

3. *Webster's Ninth New Collegiate Dictionary.* (Springfield, Mass.: Merriam Webster, Inc., 1989).

certified specialists such as soil scientists, hydrologists, environmental engineers, and biologists. In general, most appraisers are not qualified to perform this function. Therefore, to comply with the Uniform Standards of Professional Appraisal Practice (USPAP), appraisers should defer wetland delineation to specialists and acknowledge reliance on the reports or studies prepared by others.

DETERMINATION OF WETLANDS

Specialists employ two methods to identify and delineate wetland areas: off-site (desk) and on-site (field) investigations. In an off-site review, the specialist examines such information as National Wetlands Inventory Maps prepared by the U.S. Department of the Interior, soil surveys prepared by the Soil Conservation Service of the U.S. Department of Agriculture, flood maps prepared by the Flood Emergency Management Agency, and aerial photographs (to identify standing water and other features). An on-site investigation is generally only performed if sufficient information is unavailable for a desk review. In an on-site investigation, field studies determine the soils, plants, and hydrology.

A *routine* on-site investigation is performed if the site in question is less than or equal to five acres and if the vegetation appears homogeneous. An *intermediate* on-site investigation is performed if the site is greater than five acres and has diverse vegetation. A *comprehensive* on-site investigation is performed in all other cases. The complexity of the investigation increases with each step. The standard form used in *routine* evaluations appears here.

Wetland identification and boundary delineation is best left to certified specialists, but appraisers and other market participants should be familiar with the five basic types of wetland systems and know their general characteristics. These are the Marine, Estuarine, Riverine, Lacustrine, and Palustrine systems.

Marine System

In general the Marine system consists of the ocean and its extension to the high water mark along the shoreline. According to the U.S. Fish and Wildlife Service, the Marine system

> consists of the open ocean overlying the continental shelf
> and its associated high-energy coastline. Marine habitats
> are exposed to the waves and currents of the open ocean
> and the water regimes are determined primarily by the

Data Form
Routine Onsite Determination Method[1]

Field Investigator(s): _____ Date: _____

Project/Site: _____ State: _____ County: _____

Applicant/Owner: _____ Plant Community #/Name: _____

Note: If a more detailed site description is necessary, use the back of data form or a field notebook.

Do normal environmental conditions exist at the plant community? ☐ Yes ☐ No (if no, explain on back)

Has the vegetation, soils, and/or hydrology been significantly disturbed? ☐ Yes ☐ No (If yes, explain on back)

Vegetation

Dominant Plant Species	Indicator Status	Stratum	Dominant Plant Species	Indicator Status	Stratum
1.			11.		
2.			12.		
3.			13.		
4.			14.		
5.			15.		
6.			16.		
7.			17.		
8.			18.		
9.			19.		
10.			20.		

Percent of dominant species that are OBL, FACW, and/or FAC _____

Is the hydrophytic vegetation criterion met?   ☐ Yes   ☐ No

Rationale: _____

Soils

Series/phase: _____ Subgroup:[2] _____

Is the soil on the hydric soils list? ☐ Yes ☐ No ☐ Undetermined

Is the soil a Histosol? ☐ Yes   ☐ No   Histic epipedon present? ☐ Yes ☐ No

Is the soil: Mottled? ☐ Yes   ☐ No   Gleyed? ☐ Yes ☐ No

Matrix Color: _____   Mottle Colors: _____

Other hydric soil indicators: _____

Is the hydric soil criterion met? ☐ Yes ☐ No

Rationale: _____

Hydrology

Is the ground surface inundated? ☐ Yes ☐ No   Surface water depth: _____

Is the soil saturated?   ☐ Yes ☐ No

Depth to free-standing water in pit-soil probe hole: _____

List other field evidence of surface inundation or soil saturation. _____

Is the wetland hydrology criterion met?   ☐ Yes ☐ No

Rationale: _____

Jurisdictional Determination and Rationale

Is the plant community a wetland? ☐ Yes ☐ No

Rationale for jurisdictional decision: _____

[1]This data form can be used for the Hydric Soil Assessment Procedure and the Plant Community Assessment Procedure.

[2]Classification according to "Soil Taxonomy."

Source: *Federal Manual for Identifying and Delineating Wetlands* (Washington, D.C.: Federal Interagency Committee for Wetland Delineation), p. B-2.

Valuation of Wetlands

ebb and flow of ocean tides. . . . Shallow coastal indentations or bays without appreciable freshwater inflow, and coasts with exposed rocky islands that provide the mainland with little or no shelter from wind and rain are also considered part of the Marine System because they generally support typical marine biota.[4] (See Figure 1)

Estuarine System

Estuarine wetlands such as salt marshes and mangrove swamps are located in tidal areas between Marine wetlands and adjoining uplands. The Estuarine system

consists of deepwater tidal habitats and adjacent tidal wetlands that are usually semi-enclosed by land but have open, partly obstructed, or sporadic access to the open ocean, and in which ocean water is at least occasionally diluted by freshwater runoff from the land. The salinity may be periodically increased above that of the open ocean by evaporation. Along some low-energy coastlines there is appreciable dilution of sea water. Offshore areas with typical estuarine plants and animals such as red mangroves and eastern oysters are also included in the Estuarine System.[5] (See Figure 2)

Riverine System

Riverine wetlands are associated with rivers, creeks, canals, and channelized waterways. The Riverine system

includes all wetlands and deepwater habitats contained within a channel, with two exceptions: 1) wetlands dominated by trees, shrubs, persistent emergents, emergent mosses, or lichens, and 2) habitats with water containing salts in excess of 0.5 0/00 [0/00 = parts per thousand]. A channel is an open conduit either naturally or artificially created which periodically or continuously contains moving water, or which forms a connecting link between two bodies of standing water.[6] (See Figure 3)

---

4. *Classification of Wetlands and Deepwater Habitats of the United States* (Washington, D.C.: U.S. Department of the Interior, Fish and Wildlife Service, 1979), p. 4.

5. Ibid., p. 4, 8.
6. Ibid., p. 9.

FIGURE 1
Marine System

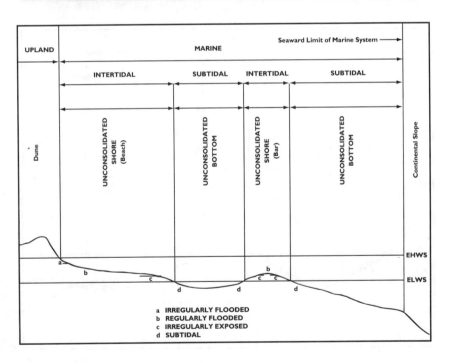

FIGURE 2
Estuarine System

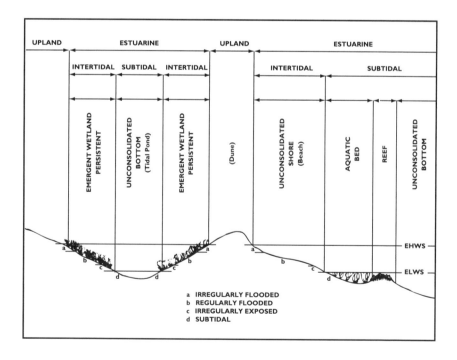

a IRREGULARLY FLOODED
b REGULARLY FLOODED
c IRREGULARLY EXPOSED
d SUBTIDAL

What Are Wetlands?

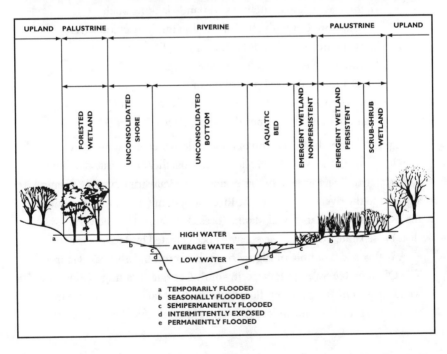

Valuation of Wetlands

Lacustrine System

Lacustrine wetlands are associated with lakes and reservoirs. The Lacustrine system

> includes wetlands and deepwater habitats with all of the following characteristics: 1) situated in a topographic depression or a dammed river channel; 2) lacking trees, shrubs, persistent emergents, emergent mosses or lichens with greater than 30% aerial coverage; and 3) total area exceeds 8 ha [hectares]. Similar wetland and deepwater habitats totaling less than 8 ha are also included in the Lacustrine System if an active wave formed or bedrock shoreline feature makes up all or part of the boundary, or if the water depth in the deepest part of the basin exceeds 2 m (6.6 feet) at low water.[7] (See Figure 4)

Palustrine System

In general Palustrine wetlands are associated with swamps and bogs separate from other wetland systems. The Palustrine system

> includes all non-tidal wetlands dominated by trees, shrubs, persistent emergents, emergent mosses or lichens. . . . It also includes wetlands lacking such vegetation, but with all of the following four characteristics: 1) area less than 8 ha [hectares]; 2) active wave formed or bedrock shoreline features lacking; 3) water depth in the deepest part of basin less than 2 m at low water; and 4) salinity due to ocean-derived salts less than 0.5 0/00 [parts per thousand].[8] (See Figure 5)

Figure 6 shows the interaction of these various wetland systems. As can be seen in Figure 6, the systems are often linked together and create one large ecologic, hydrologic system. In this example, the Palustrine system (swamp adjacent to the river) is linked to the Riverine system (river), which is linked to the Estuarine (salt marsh) and Marine (ocean) systems. The Lacustrine system (lake) is separate from the others, having been created by a reservoir dam.

A real world example of Figure 6 can be found in southeast Georgia. The Okefenokee Swamp, located in south Georgia, is a huge Palustrine wetland system. This swamp flows into the St. Mary's River (Riverine system), which then meanders easterly to the salt marshes (Estuarine

---

7. Ibid., p. 11.   8. Ibid., p. 12.

FIGURE 4
Lacustrine System

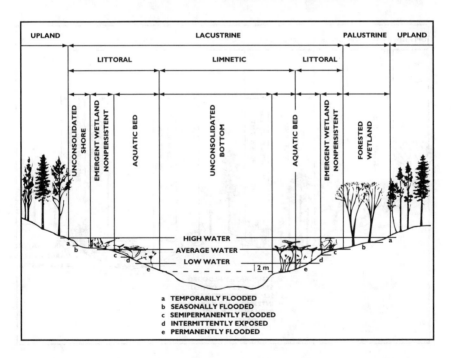

FIGURE 5
Palustrine System

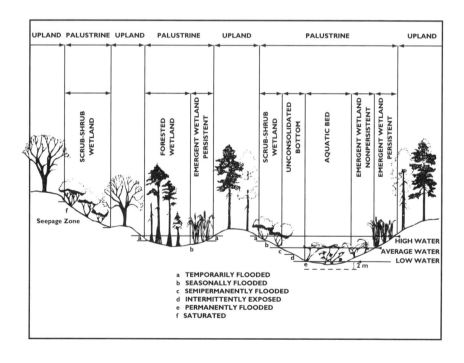

UPLAND PALUSTRINE UPLAND PALUSTRINE UPLAND PALUSTRINE UPLAND

SCRUB-SHRUB WETLAND

FORESTED WETLAND

EMERGENT WETLAND PERSISTENT

SCRUB-SHRUB WETLAND

UNCONSOLIDATED BOTTOM

AQUATIC BED

EMERGENT WETLAND NONPERSISTENT

EMERGENT WETLAND PERSISTENT

f
Seepage Zone

HIGH WATER
AVERAGE WATER
LOW WATER

2 m

a TEMPORARILY FLOODED
b SEASONALLY FLOODED
c SEMIPERMANENTLY FLOODED
d INTERMITTENTLY EXPOSED
e PERMANENTLY FLOODED
f SATURATED

What Are Wetlands?

FIGURE 6

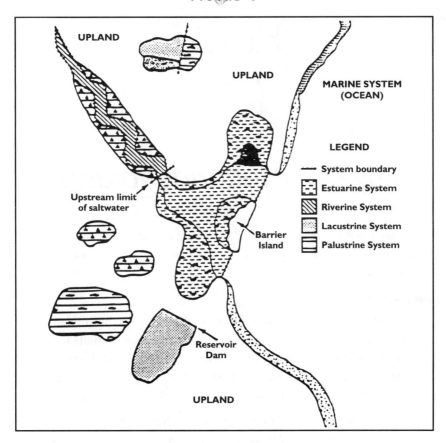

system) along the coast, and eventually flows into the Atlantic Ocean (Marine system). Similar examples can be found in most coastal areas.

SUMMARY

Wetlands are characterized by distinct soils, vegetation, and hydrology. Their identification and delineation should be left to specialists. If the appraiser is not such a specialist, he or she should avoid making wetland delineations and rely on the expertise of others to comply with USPAP. This must be clearly disclosed in any appraisal report.

The five basic types of wetland systems are the Marine, Estuarine, Riverine, Lacustrine, and Palustrine systems. Appraisers and other market participants should be aware of these systems and know their general characteristics. This is especially important when establishing like units of comparison for analysis purposes.

Chapter 2

Why Are Wetlands Valuable?

Wetlands are valuable because of their unique functions and characteristics. In this chapter, the physical, legal, social and economic aspects of wetlands are examined.

PHYSICAL

Wetlands serve many physical functions critical to the health and maintenance of our ecosystem. In particular, they are habitats for wildlife and fisheries, they provide water emergence and filtering systems, they serve adjoining uplands as buffers against flooding, and they limit the effects of erosion. Regarding habitat functions,

> 80% of the Nation's coastal fisheries are dependent on wetlands for spawning, nursery areas, and food sources. Both coastal and inland wetlands provide essential breeding, nesting, feeding, and predator escape habitats for millions of waterfowl, other birds, mammals and reptiles. Well over one third of the 564 plants and animal species listed as threatened or endangered in the United States utilize wetland habitats during some portion of their life cycle.[1]

Wetlands "also provide essential nesting, migratory, and wintering areas for more than 50% of the Nation's migratory birds."[2] These functions are vitally important to the ecosystem and make possible the over $10 billion fishing industry and the over $40 billion hunting industry.

---

1. *Wetland Stewardship,* U.S. Department of the Interior, April 1992.

2. T.E. Dahl and C.E. Johnson, *Status and Trends of Wetlands in the Coterminous United States, Mid-1970s to Mid-1980s* (Washington, D.C.: U.S. Department of the Interior, Fish and Wildlife Service, 1991).

Regarding hydrological functions, wetlands can serve as water emitters and/or drains, depending on the type of system. For example, wetlands that include natural springs emit water from the underlying aquifer into a river, stream, or other body of water, even the ocean, and thus serve as a source of fresh water. Many cities and towns make emitting wetlands their primary source of public water. Other wetlands serve as drains that move rain water and other runoff down through the ground and into the underlying aquifer, filtering and recharging the water in the process. According to the U.S. Fish and Wildlife Service, "the groundwater discharge function of wetlands [movement of groundwater into surface water, e.g., springs] is recognized as being more important than the groundwater recharge function [movement of surface water into ground-water aquifers]."[3]

In addition to serving as emitters and drains, wetlands also serve as pollution filters strategically placed between uplands and open water sources. For example, rain and other water runoff draining from uplands to open waters typically must pass through a wetland system. As this runoff passes through the wetland, vegetation, soils, and other components collect particulants (including pollutants), thus preventing them from entering the open waters. The ability of a wetland to perform this function is finite, however, and is related to its size, type, and quality and to the amount of runoff being processed.

Wetlands also serve as the first line of defense against flood waters. In this capacity, they absorb and slow rising water levels. The following underscores the importance of wetlands as flood buffers:

> Many authors cite the Corps of Engineers' 1972 study of the Charles and Neponset River watersheds in Massachusetts as a prime example of the socioeconomic values associated with protecting wetlands to maximize flood control. In this study, the Corps estimated that loss of the 8,423 acres of wetlands within the basin would result in annual flood damages of over $17,000,000.[4]

The recent flood damage along the Mississippi River is good testimony of the devastation that can result when wetlands are destroyed by agricultural conversion and development uses.

---

3. *National Wetlands Priority Conservation Plan*, U.S. Department of the Interior, Fish and Wildlife Service, April 1989, p. 19.

4. Ibid., p. 21.

Wetlands also serve an important role in reducing erosion. The hydrophytic vegetation and hydric soils form the backbone against which uplands are sustained. If such vegetation and soils are removed, erosion in the form of gullies and ditches can result. If erosion is severe, soil runoff can clog a channel over time necessitating expensive dredging operations.

LEGAL

As discussed in Chapter 1, the federal government historically encouraged wetland depletion, and statistics show this influence contributed to the destruction and degradation of over 50% of the nation's wetlands. More recently, in response to discoveries of the important functions of wetlands and their alarming demise, the government has reversed its position. Various legislative and regulatory changes have occurred to protect wetlands as an important and threatened natural resource.

The government reversal began with amendments to the Federal Water Pollution Control Act in 1972, 1977, and 1986 (Clean Water Act). As a result of these amendments, a permit from the U.S. Army Corps of Engineers must be obtained to dredge (demuck or drain) or fill (cover over and develop) wetlands. The Corps has three general guidelines in evaluating applications for such permits, which can be summarized as follows (parenthetic phrases have been added for clarification):

1. Avoidance (of impact when possible)
2. Minimization of impact (if avoidance is not possible)
3. Compensation (mitigation of impact if the impact is severe).[5]

The Clean Water Act left many loopholes for noncompliance, however, and it had little effect on the decline of wetlands. As a result, 2.6 million more wetland acres were lost from the 1970s to the 1980s, with 54% of this loss due to agricultural conversion.[6]

To close some of the loopholes in the Clean Water Act, Congress added a wetlands provision in the Food Security Act of 1985, commonly known as the Swampbuster Provision. This provision made any farmer who converts wetlands to agricultural uses ineligible for price supports, payments, crop insurance, or government-insured loans. This action threatened a severe financial blow to farmers involved in wetland conversion. Yet, as a result of poor enforcement by agricultural agencies, the provision had little effect on wetland decline.[7]

---

5. William Bunkley and Charles Edmonds III, Ph.D., "Appraising Wetlands," The Appraisal Journal (January 1992), p. 109.

6. Dahl and Johnson.

7. Bunkley and Edmonds, p. 107.

Legislative activity concerning wetlands continued with passage of the Emergency Wetlands Resources Act of 1986. The goal of this legislation was "to promote the conservation of migratory waterfowl and to offset or prevent the serious loss of wetlands by the acquisition of wetlands and other essential habitat, and for other purposes."[8] The Act intensified "efforts to protect the wetlands of the Nation through acquisition in fee, easements, or other interests and methods."[9] Therefore, in addition to regulation, this Act enabled the federal government to increase its protection of wetlands through buying property rights (fee simple and easements).

In 1989, the North American Wetlands Conservation Act was enacted, which built on previous legislation to "conserve North American wetland ecosystems and waterfowl and the other migratory birds and fish and wildlife that depend on such habitats."[10] This Act is interesting because it expands the focus of legislation from strictly wetlands to wetland ecosystems. The Act recognized that wetlands are a vital part of an entire ecosystem, which also includes uplands. As a result, acquisition and conservation efforts were broadened to include adjoining upland acres. To provide guidance to federal agencies in this regard, the Report of the U.S. Senate Committee on Environmental and Public Works (U.S. Senate, September 16, 1986) stated that

> Acquisition should be limited to those purchases of fee title or easements of wetlands and associated upland areas that contribute appreciably to the long term preservation of such wetlands and associated populations of fish, wildlife, and plants. Acquisition of uplands areas adjacent to wetlands is often essential to maintaining the values of those wetlands. . . .[11]

One example of a government acquisition occurred on September 10, 1992. The U.S. government purchased the fee simple estate in Broward Island for $835,000. The island, which actually consists of four smaller islands, is located within the Estuarine wetland system along the Nassau River, separating Duval and Nassau counties in northeast Florida. The islands contained an entire ecosystem of 293.74 acres, 33 acres of which were wetlands (11%) and 260.74 acres (89%) were uplands. The islands were purchased as an addition to the Timucuan Ecological and Historical Preserve managed by the National Park Service of the U.S. Department of the Interior.

---

8. Emergency Wetlands Resources Act. Public Law 99-645. 100 STAT. 3582.

9. Ibid.

10. North American Wetlands Conservation Act. Public Law 101-233. 103 STAT. 1968.

11. National Wetlands Priority Conservation Plan, p. 25.

Another wetlands restoration project directed at conservation is along the Trinity River in Texas. As described in the Report of the Interagency Committee on Wetlands Restoration and Creation,

> Prior landowners had installed a levy to remove the threat of overbank flooding of the Trinity River. This levee isolates the site from nearly 3,000 acres of Trinity River bottomlands. . . . Hydrology has been restored through the installation of low head levees, complete with water control structures. . . . The restored wetlands have provided valuable habitat for a wide range of shorebirds, wading birds, reptiles and amphibians, resident and migratory waterfowl, and other wetland associated wildlife.[12]

Government regulation of wetlands peaked on August 9, 1991, when President Bush issued the Executive Order containing the "no net loss" goal. The order was a line drawn in sand with respect to wetland decline: its implementation does not necessarily preclude the destruction or degradation of wetlands (though such is discouraged) but governs their net loss. In other words, it recognizes that some development will adversely impact wetlands. When such impacts are unavoidable, a developer must create, restore, or enhance other wetlands to take the place of those to be affected. In this manner, the no net loss goal can be achieved.

The legal status of wetlands remains dynamic. As of early 1995 new legislation that could further impact wetlands was introduced in Congress. One bill, the Private Property Rights Act of 1995 (H.R. 925), would govern cases of regulatory takings and provide compensation to private property owners whose property value declines as a result of government actions. Under this proposal, if wetland regulations resulted in a property value loss, the owner would be entitled to compensation. Congress is also examining the Clean Water Act. One proposed change would redefine wetlands to include only large, environmentally significant systems and allow small, isolated wetlands to be developed or converted to alternate uses without the need for permits or mitigation. The Clinton Administration has voiced opposition to both pieces of proposed legislation.

---

12. *A National Program for Wetlands Restoration and Creation,* Report of the Interagency Committee on Wetlands Restoration and Creation, August 1992.

ECONOMIC AND SOCIAL

Wetlands provide economic and social benefits, which are evident when analyzing the fishing, hunting, leisure, and real estate industries. Wetlands serve as breeding, nursery, and predatory escape habitats for wildlife, and most commercially harvestable marine life spends some stage of its life cycle in Estuarine and Marine wetlands. In addition, freshwater wetlands provide habitat for inland harvestable wildlife such as alligators, muskrats, and certain waterfowl. The skins, meat, and feathers of such wildlife plus the hunting activity wildlife generates are important sources of income for many communities. According to the Fish and Wildlife Service, "A commercial marine fisheries harvest valued at $10 billion annually (1986 dollars) provides one economic measure of the significance of coastal wetlands."[13]

The Fish and Wildlife Service also reports that "17.4 million hunters spent about $5.6 billion on supplies, lodging, transportation, and other related expenses in 1980. Of these totals, 5.3 million hunted waterfowl, spending about $640 million. In total, fish and wildlife related recreation in 1980 was a $41 billion industry based largely on wetlands resources."[14]

Wetlands also influence real estate development trends. Typically, residential, lodging, and restaurant developments prefer scenic locations such as are found along Marine, Estuarine, Riverine, and Lacustrine wetland systems when possible. Such locations often provide a panoramic open view, which is generally pleasing and attractive to consumers. The economic benefit of properties enjoying such views is easily measured in the market. For example, properties with views and development buffers created by wetlands typically sell for more than those lacking such features, all other things being equal. As such, it is clearly evident that wetlands add value to adjoining developable uplands. However, wetlands generally have significantly less value than uplands due to their highly restricted development potential.

SUMMARY

Wetlands are valuable because of their unique physical, legal, economic and social characteristics. These characteristics should be recognized and thoroughly investigated by appraisers in the market valuation process.

---

13. *National Wetlands Priority Conservation Plan*, p. 18.
14. Ibid., p. 22.

Chapter 3

Controversial Values

A wide range of values can be found in appraisals of wetlands, and this creates a great deal of controversy. Typically real estate appraisers estimate lower wetland values than do environmentalists, academicians, and certain government agencies. Who is right? This chapter examines the controversy to assist appraisers in defending their opinions and help them understand the opinions of others.

The primary reason real estate appraisers and environmentalists arrive at different value opinions of wetlands is because the two parties are often solving different valuation problems. Specifically, an analysis of the methods used by the two reveals that real estate appraisers typically appraise the *market value of an interest* in a wetland, whereas environmentalists typically consider the *total economic value* of a wetland, *irrespective of interest.* Since the two problems are different, it should come as no surprise that the two solutions are different. Addressing this point, Anderson and Rockel write:

> Market prices measure only a part of willingness to pay. Consumers who buy goods and services pay the market price. This means that the commodity is worth to them at least the market price; some individuals may well have been willing to pay more had that been necessary. This benefit that individuals obtain in excess of market price is termed "consumer surplus." Thus, the total value of a good or service is the summation of market price and consumer surplus. . . . While wetland acreage purchases and sales are relatively common, these prices will understate the social value of wetlands because private

owners are unable to capture many of the benefits. Thus, market transactions for wetlands do not provide a good guide to [total economic] wetlands values.[1]

MARKET VALUE AND PROPERTY RIGHTS

Appraisers are generally requested to estimate the *market value* of a *specified interest* in real estate such as the market value of the fee simple estate. Market value is the most probable price of a property subject to certain restrictions such as well-informed or well-advised parties, a reasonable time for exposure in the open market, a fair sale without undue duress, cash or equivalent financing, etc. Property rights are also subject to restrictions, i.e., the fee simple estate is subject to the governmental powers of taxation, eminent domain, police power, and escheat. In contrast the only factor restricting total economic value is society's ability to recognize the benefits of a property and translate those benefits into an indication of value.

To expound on this point, it should be recognized that the law acknowledges both the private property owner and the public as a whole are beneficiaries of wetland functions. "Under the law, the right of individuals to own and use land for material gain is maintained, while the right of all people to use the land is protected."[2] Therefore, two basic sets of interests are present, which may be expressed as an equation:

Private interest value
+ Public interest value
———————————
= Total economic value

As previously discussed, laws governing wetlands are vast and effectively protect and conserve wetland functions for the welfare of the public, creating public interest value. If public interest has a positive value, then the resultant value to the private property owner will be less than the total economic value of the wetland as a whole. Appraisers typically should not include public interest value in their estimate of the value of a private property owner's interest.

The following series of graphs will help illustrate this often confusing concept. Figure 1 represents the total value of a parcel of real estate. It includes 100% of the private and public interests in a property.

1. Robert Anderson and Mark Rockel, *Economic Valuation of Wetlands*, Discussion Paper 65 (American Petroleum Institute, April 1991), p. 14, 16.

2. Appraisal Institute, *The Appraisal of Real Estate*, 10th edition (Chicago: Appraisal Institute, 1992), p. 4.

Figure 2 shows how value can be divided into both private and public interests. In this case, the private interest enjoys the lion's share of rights and benefits. An example of such a case is the fee simple interest in a 100% developable, residentially zoned homesite. The public interest is minor and is created by zoning and civil laws which restrict some of the legally permissible uses of the property, i.e., the zoning may limit permissible uses, local laws may limit noise emissions and certain activities, etc. Figure 3 represents a wetland scenario. In this case, government laws and regulations take the lion's share of value from the private property owner to protect the public interest. Therefore, when appraisers are estimating the value of a private property owner's interest in a wetland, they are estimating the value of a slice of the overall value pie. Appraisers should recognize this division of property right values.

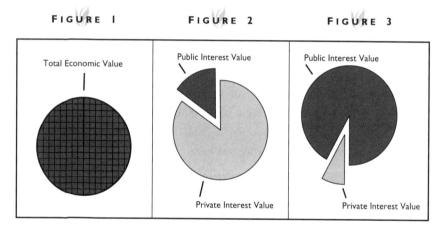

FIGURE 1   FIGURE 2   FIGURE 3

Total Economic Value

Public Interest Value

Public Interest Value

Private Interest Value

Private Interest Value

ESTIMATING TOTAL ECONOMIC VALUE

The valuation approaches that appraisers should employ when estimating the market value of an interest in a wetland are discussed in the next chapter. However, to help appraisers understand the position of others, it is useful to examine the valuation approaches being developed by environmentalists and academicians to estimate the total economic value of a wetland. Most of these approaches are based on the sum of the values of the various functions that wetlands provide. Specifically, the value of a wetland equals the sum of such functions as wildlife nursery, wildlife habitat, water emergence, flood buffer, and water filtration. Various methods have been proposed to estimate these values. For example, Dr. Frederick Bell at Florida State University has performed extensive research on the economic value of wetlands, and he proposes the following method to estimate the nursery value of Estuarine wetlands:

> The marginal product of an acre of estuarine wetland, holding all other factors constant, is valued by the price people are willing to pay for the fishery product. Since this marginal product will flow into perpetuity, at least from a theoretical point of view, one may estimate the capitalized value of this flow by dividing by the discount rate.[3]

Dr. Bell has developed a variation of the income approach to estimate the economic value of the nursery benefits of Estuarine wetlands. However, he understands that there are different valuation problems and recognizes the difference between the value of a wetland as a whole versus the value of a private property owner's interest in that wetland. "Private owners could restrict access and charge fees for some (not all) of the natural wetland functions. For example, hunting clubs often rent wetlands for hunting waterfowl or fur bearing animals. But. . . *a private owner can seldom, if ever, charge for the multiplicity of wetland services such as those derived from the food chain.*"[4] (Emphasis added.)

In summary, the market value of a private property owner's interest in a wetland may differ from the total economic value of the wetland.

While the preceding has attempted to provide some insight into the diverging opinions of wetland values, the following hypothetical examples may help to clarify the issue further.

Example 1

Assume a local government is seeking to acquire 200 acres of wetlands from a rancher for airport expansion purposes. However, the rancher doesn't want to sell. As a result, the local government decides to exercise its powers of eminent domain and condemn the property for public use. According to law, the government must pay the rancher just compensation. To decide this compensation, the two parties go to court and argue their estimates of value before a judge and jury.

The local government hires an appraiser who, based on sales and other information, estimates the market value of the rancher's fee simple interest in the wetlands at $400 per acre. The rancher, on the other hand, hires a university-based researcher who estimates the total economic

---

3. Frederick W. Bell, *Application of Wetland Valuation Theory to Florida Fisheries* (Florida State University, June 1989), p. iii.

4. Bell, p. 28.

value of the wetlands at $3,000 per acre. The methods used by both the appraiser and the university researcher are valid and accurate. However, the resulting difference is vast, for one value is associated with the market value of the rancher's interest whereas the other value is associated with the total economic value of the wetland, irrespective of interest.

The judge, a former real estate attorney, rules on the matter and throws out the testimony of the university researcher. In this case, he declares it would be improper for the public to pay for rights they already enjoy and technically own (as a result of a regulatory taking). The university researcher's numbers weren't wrong, they just weren't relevant to the problem. Therefore, the rancher is awarded $400 per acre as just compensation for the taking.

A variation of this example can be found in a recent court decision. According to Crookshank

> *Lucas v. South Carolina Coastal Council* (112 S. Ct. 2886, 22 ELR 21104, 1992) established that if a government regulatory action destroys "all economically beneficial or productive use" of privately owned property, the action can represent a compensable taking. This means that the government must compensate the property owner for the losses incurred as a result of the regulatory action. In wetlands regulation, this means that if an owner of a wetland is denied a permit to develop his wetland, and the land has no other economically beneficial or productive use, he can bring action against the regulatory agency for compensation.[5]

This lawsuit occurred in the wake of Hurricane Hugo and the devastation it caused to the South Carolina coast. A coastal property owner sought to rebuild after the storm but was denied this right by government. The court ruled that if regulations effectively take away a property owner's rights, then this is a taking and compensation must be paid. The decision recognizes the division of property rights and is expected to spur a wave of new lawsuits against the government for compensation for such regulatory takings.

---

5. Steven Crookshank, *Air Emissions Banking and Trading: Analysis and Implications for Wetland Mitigation Banking*, Research Study #074 (American Petroleum Institute, February 1994), p. 5.

Example 2

In contrast to the above example, assume an oil tanker runs aground on a reef near an exclusive resort area, discharging millions of gallons of crude into the ocean. This crude washes onto the beach and severely damages an important Estuarine wetland area, which is part of a 500-acre coastal resort owned by a wealthy businessman. In response to this disaster, the state and the property owner jointly sue the oil company for compensation.

The property owner hires an appraiser and the state assembles a team of scientists, and both argue their case before the court. The property owner's appraiser estimates a market value of the fee simple interest at $300 per acre, and the state's experts argue a total economic wetland value of $3,500 per acre. Both valuation methods are valid yet yield different amounts. The court rules the spill affected both the private property owner and the public (as represented by the state) and awards a value of $3,500 per acre to be allocated between the property owner and state per their interests.

In this example, the effect on total economic value is pertinent, for both the private property owner and the public were affected. Therefore, it can be seen that a solid understanding of the appraisal problem is critical to the market valuation of wetlands.

SUMMARY

Appraising wetlands is controversial due to varying opinions and methods of valuation. The reason for most of this controversy is a misunderstanding of the appraisal problem. Some key points in defining the appraisal problem are understanding the type of value being appraised, knowing the property rights being appraised, and identifying the appropriate valuation approach. If the problem is to estimate the market value of an interest in a wetland such as the fee simple estate, then the techniques discussed in Chapter 4 can be employed. However, if the problem is to estimate the total economic value of a wetland, then an appraiser may want to defer to other individuals who can estimate the value of the various wetland functions.

Chapter 4

The Valuation Approaches

The three approaches developed to estimate the market value of real property should all be used, as data permits, when appraising wetlands. Applications of the sales comparison, cost, and income approaches to the valuation of wetlands are discussed in this chapter.

SALES COMPARISON APPROACH

By far the most common and generally accepted approach is the sales comparison approach, which is based on the principle of substitution. The underlying assumption is that market value is a function of the price to acquire an alternative property offering similar features. It is assumed a buyer realizes that one property can generally, although not exactly, be substituted for another and that differences in properties are negotiated on the basis of price. In this approach, recent sales of comparable properties in an area are identified and researched and then compared to the property being appraised. Differences between the comparable sales and the subject property are then recognized, and adjustments for these differences are made based on market-extracted information. The adjusted sales are then used as indicators of market value.

The advantage of the sales comparison approach is that it permits direct comparison of a property under analysis to actual market transactions; it is also the most easily understood approach. However, it has several limitations, one of which relates to the fact that no two properties are exactly the same and a perfect substitute is not possible. Other limitations result from the inefficiencies and imperfections of the real estate market, which make consistent market-extracted adjustments difficult if not impossible. As such, appraisers must always apply sound judgment in the valuation process, and this judgment is a function of the appraiser's experience and education.

It is unusual to appraise a property consisting of 100% wetlands. Instead, most properties consist of a mixture of both wetlands and uplands. Therefore, two different property types are often involved. Uplands typically sell for more than wetlands because of their greater development potential. The appraiser, therefore, should be sure to research the percentage of uplands to wetlands for both the subject property and the comparable sales. This said, there are basically three unique applications of the sales comparison approach to the market valuation of an interest in wetlands. These include a *whole-to-whole* analysis, a *sum-of-the-parts* analysis, and a *residual* analysis.

Whole-to-Whole Analysis

In a whole-to-whole analysis sales are compared to the subject property on a gross acreage basis. The key to proper application of this method is to identify sales having wetland ratios similar to the property being appraised or to be able to adjust for differences in wetland ratios based on market-extracted data. Otherwise, the appraiser will not be comparing like units.

In addition, like systems should be compared, for some systems sell for more than others. For example, a coastal property containing Estuarine wetlands associated with a salt marsh may sell for more than an inland property containing Palustrine wetlands associated with a swamp, all other things being equal. The valuation formula in a whole-to-whole analysis is

$$V_o = UV_g \times U_g$$

where

V_o = Market value of the (whole) property
UV_g = Unit value per gross acre
U_g = Units of gross acres

Example

Assume the fee simple estate in a 1,000-acre tract of vacant land is being appraised. The tract consists of 80% Palustrine wetlands associated with a swamp, and the other 20% consists of forested uplands. Four comparable sales have been identified for analysis.

Sale	Date	Price	Acres	Percent Wetlands	Price/ Acre
1	01/15/94	$1,000,000	1,000	80%	$1,000
2	03/27/94	$1,017,500	925	80%	$1,100
3	12/01/94	$1,500,000	1,000	60%	$1,500
4	03/10/95	$1,350,000	1,050	70%	$1,286

Assume confirmation reveals all the sales involve the transfer of a fee simple estate and were arm's-length transactions based on all-cash financing. Therefore, no adjustments are required for property rights conveyed, conditions of sale, or financing. Also assume market conditions have been flat, all the sales contain Palustrine wetlands associated with a swamp and have equivalent physical features (except for size), and none contains harvestable vegetation. Therefore, no adjustments are needed for variations in market conditions, wetland systems, physical features, or other values.

As can be seen, the sales are fairly comparable in size, ranging from 925 to 1,050 acres, and they yield unit prices ranging from $1,000 to $1,500 per gross acre. An analysis of these sales reveals two obvious adjustments. The first is for differing wetland ratios, and the second is for differing sizes.

To begin, the adjustment for differing wetland ratios can be extracted by comparing Sales 1 and 3. Sale 1 contains 1,000 acres with a wetland ratio of 80% and sold at a unit price of $1,000 per gross acre. In contrast, Sale 3 contains 1,000 acres but had a wetland ratio of only 60%, and it sold for $1,500 per gross acre. Sales 1 and 3 are comparable in every way except for wetland ratio, and since Sale 3 sold for more than Sale 1, it is assumed this is due to its lower percentage of wetlands. Based on this example, an appraiser can deduce that a higher percentage of wetlands on a tract lowers its gross unit value and, conversely, a lower percentage of wetlands increases its gross unit value. This assumption is logical and reasonable, for wetlands have a much lower development and use potential compared to uplands. Evidence of this adjustment is consistent among the sales.

The second evident adjustment is for size, which can be extracted by comparing Sales 1 and 2. Sale 1 contains 1,000 acres with 80% wetlands and sold for $1,000 per gross acre. Sale 2 contains 925 acres with a wetland ratio of 80% and sold for $1,100 per gross acre. Both Sale 1 and Sale 2 are comparable in every way except size, and since Sale 2 sold for more than Sale 1, this can be attributed to its smaller size. In this example, the appraiser deduces that size influences unit values. This adjustment is logical and reasonable, and consistent with economies of scale. All of the sales show evidence of this adjustment.

After comparing the sales on a whole-to-whole basis, Sale 1 appears to be the best indicator of value for the property being appraised because of its comparable size and wetlands ratio. Therefore, in this example, the unit value of the subject property is estimated at $1,000 per gross acre. Applying this unit value in the valuation formula yields the following solution:

$$V_o = UV_g \times U_g$$
$$V_o = \$1{,}000/\text{acre} \times 1{,}000 \text{ acres}$$
$$V_o = \$1{,}000{,}000$$

In a whole-to-whole analysis it is important that comparable sales share similar wetlands ratios and similar wetland system characteristics or that adjustments be extracted from the market via a paired sales analysis to compensate for differences in these characteristics. The unit of comparison in a whole-to-whole analysis is the price per *gross* acre.

Sum-of-the-parts analysis

In a *sum-of-the-parts* analysis, the market values of the wetlands and the uplands are estimated separately, then added together to estimate the market value of the property as a whole. This method is based on the following formula:

$$V_o = V_w + V_u$$

where

V_o = Market value of the (whole) property
V_w = Market value of the wetlands
V_u = Market value of the uplands

or

$$V_o = (UV_w \times U_w) + (UV_u \times U_u)$$

where

V_o = Market value of the (whole) property
UV_w = Unit value of the wetlands
U_w = Units of wetlands
UV_u = Unit value of the uplands
U_u = Units of uplands

As in a whole-to-whole analysis, the key to a sum-of-the-parts analysis is identifying comparable data so that like units can be compared. Therefore, wetland sales should be used to estimate the market value of the wetlands being appraised, and upland sales should be used to estimate the market value of the uplands being appraised.

Example

Assume the same property as in the previous example. As stated, the tract contains 1,000 acres with 80% Palustrine wetlands associated with a swamp and the remaining 20%, forested uplands. Therefore, the property contains 800 acres of wetlands and 200 acres of uplands. The following sales have been identified and grouped into upland and wetland categories.

WETLANDS					
Sale	Date	Price	Acres	Percent Wetlands	Price/ Acre
1	02/14/94	$500,000	1,000	100%	$500
2	07/10/94	$350,000	750	100%	$467
3	12/20/94	$425,000	900	95%	$472
UPLANDS					
4	08/21/95	$545,000	170	1%	$3,206
5	11/15/93	$758,000	240	5%	$3,158
6	01/21/94	$700,000	225	3%	$3,111

Again, assume all the sales were confirmed to involve the transfer of a fee simple estate, were arm's-length transactions, and were based on all-cash financing. Also, assume all wetlands are part of a Palustrine system associated with a swamp, the sales have equivalent physical features, and none of the sales contains harvestable vegetation.

As can be seen, the wetland sales range from $472 to $500 per gross acre and the upland sales range from $3,111 to $3,206 per gross acre. In these cases wetlands sold for roughly 15% to 20% of the value of uplands. Such ratios are common. After comparing the wetland and upland sales to the component parts of the subject, the market value of the wetlands is estimated to be $500 per acre and the market value of the uplands to be $3,200 per acre. The value of the whole is estimated via a sum-of-the-parts as follows:

$$V_o = (UV_w \times U_w) + (UV_u \times U_u)$$
$$V_o = (\$500/acre \times 800\ acres) + (\$3,200/acre \times 200\ acres)$$
$$V_o = \$400,000 + \$640,000$$
$$V_o = \$1,040,000$$

It is recognized that it is difficult to identify sales in the market consisting completely of one land type. Therefore, if only minor amounts, say less than 10%, of another land type are present, then such sales can be used if small adjustments are made.

Residual Analysis

If sales of only one component part are available, then a mixture of the whole-to-whole and sum-of-the-parts methods can be applied in a *residual analysis*. In this case, if two of the components of the formula are known, then the third component can be solved using algebra. If sales of overall properties with like wetland ratios have been identified for a whole-to-whole analysis, and sales of all uplands have been identified but no sales of wetlands can be found, then algebra can be used to solve for the residual wetland component. For example, if sales of comparable parcels on a gross acre and an upland basis are identified but no wetland sales are found, the value of the wetlands can be estimated using the following formula:

$$V_o = V_w + V_u$$

Subtracting the value of the uplands
from both sides of the equation yields

$$V_w = V_o - V_u$$

or

$$V_w = (UV_o \times U_o) - (UV_u \times U_u)$$

Example

Assume a 500-acre parcel is being appraised that contains a 50% wetland ratio. No wetland sales can be found. Assume a whole-to-whole analysis yields a value estimate of $2,000 per gross acre, or $1,000,000, and plenty of comparable upland sales in the 250-acre size range can be found with an indicated value of $3,000 per acre. Applying these inputs to the above formula yields the following solution:

$$V_w = (UV_o \times U_o) - (UV_u \times U_u)$$

$$V_w = (\$2,000/\text{acre} \times 500 \text{ acres}) - (\$3,000/\text{acre} \times 250 \text{ acres})$$
$$= \$1,000,000 - \$750,000$$
$$= \$250,000$$

Dividing the wetland value of $250,000 by the amount of wetlands, or 250 acres (50% of 500 acres), yields a unit value of $1,000 per wetland acre.

The residual method is useful when the value of the component land types must be estimated and there is insufficient information to make a sum-of-the-parts method viable by itself.

COST APPROACH

The cost approach, like the sales comparison approach, is based on the economic principle of substitution. In this case, however, the premise is that the value of a property should be no more than the cost to purchase and construct a substitute property. In the cost approach, costs are the unit of comparison in estimating value, and there are two types of costs: reproduction costs and replacement costs. Each term can be defined as follows (for clarity, the word *wetlands* replaces the word *building*):

> **Reproduction cost:** the estimated cost to construct, at current prices as of the effective date of appraisal, an exact duplicate or replica of the [wetland] being appraised, using the same materials, construction standards, design, layout, and quality of workmanship and embodying all the deficiencies, superadequacies, and obsolescence of the subject [wetland].[1]

> **Replacement cost:** the estimated cost to construct, at current prices as of the effective appraisal date, a [wetland] with utility equivalent to the [wetland] being appraised, using modern materials and current standards, design, and layout.[2]

As can be seen, reproduction cost refers to an exact duplicate, whereas replacement cost refers to a functional equivalent. This is an important difference, for it is impossible to construct an exact duplicate of a wetland. However, scientific discoveries and technological advances have enabled engineers to functionally create and restore wetlands. Thus, replacement cost is the appropriate unit of comparison with respect to wetland valuation.

In general, the only time the cost approach is applicable in the market valuation of wetlands is in the appraisal of prime properties for development that are subject to mitigation (i.e., compensation of wetland losses resulting from development). This approach has become applicable as a result of the government's intervention in the marketplace and recent scientific discoveries.

Through intervention in the marketplace, the government has established a system in which the creation and restoration of wetlands are

---

1. Appraisal Institute, *The Dictionary of Real Estate Appraisal,* 3rd edition (Chicago: Appraisal Institute, 1993), p. 304.

2. Ibid., p. 303.

sometimes required as a form of mitigation. This system has arisen in response to the government's stated goal of "no net loss" of wetlands. If a wetland is destroyed in the development of a site, then government agencies may require an additional wetland be created, restored, enhanced, or preserved to take its place.

Wetland creation is the most intense form of mitigation and involves the conversion of uplands into wetlands. The process varies depending on the characteristics of the land being converted. Typically it includes removing existing topsoil and replacing it with hydric soils, removing existing vegetation and replacing it with hydrophytic vegetation, and altering existing hydrology to create anaerobic conditions. If successful, created wetlands can be functionally equivalent to natural wetlands and eventually draw birds and other wildlife.

The series of photographs on these pages depicts a wetland creation project. This project was associated with the mitigation of wetland impacts for the development of a new regional mall in Jacksonville, Florida.

Figure 1 shows uplands that have had existing topsoil removed and an indentation in elevation created for hydrologic purposes. Figure 2 shows the same site after the hydrology has been altered, creating a pond of

FIGURE I

Valuation of Wetlands

FIGURE 2

FIGURE 3

The Valuation Approaches

water. As can be seen, hydrophytic vegetation has been planted. In Figure 3 the created wetland is mature and thriving. In particular, note how the hydrophytic vegetation has spread across the wetland.

Wetland creation is expensive, and costs may run from $10,000 to $60,000 per acre exclusive of land costs. Therefore, creation is usually only feasible when mitigating the development of prime properties.

Wetlands can also be restored or enhanced. Degraded wetlands can be restored to their prior natural state, usually by removing or rerouting dikes, levees, drainage ditches, or other barriers to restore natural hydrology. However, if extensive restoration is needed, the cost can nearly equal that of creation.

Wetland enhancement is similar to restoration but is not as involved. A degraded wetland is enhanced by improving one or more wetland functions. For example, invasive plant species may be removed to ensure hydrophytic vegetation can thrive.

The lowest form of mitigation is preservation. In these cases one wetland area is preserved to compensate for other wetlands removed through development. The cost is basically that of acquiring the needed number of wetland acres to be preserved. This is the least favorite form of mitigation among government agencies, for it does not meet the goal of "no net loss." In such cases, the mitigation ratio is generally very high.

For example, assume development of an office complex will adversely impact two acres of wetlands. One mitigation alternative is preservation. Instead of asking the developer to create two additional wetland acres to replace those impacted, regulators may ask him to buy good quality wetlands off-site at a ratio of 50 to 1 and dedicate those wetlands via a conservation easement for preservation/conservation. In this way, regulators can preserve 100 acres of good quality wetlands (50 x 2 = 100) instead of opting for the creation of two acres of isolated wetlands.

In addition to the hard costs of moving dirt, planting vegetation, and maintaining a new wetland over time, several soft costs are also involved in wetland creation, restoration, and enhancement. The most common include fees for environmental engineering, design and project oversight, and attorneys and public agency permits and other legal compliance. Other soft costs include those associated with the lengthy conversion and monitoring/maintenance process.

When estimating wetland replacement costs, there are two primary sources of cost information: environmental engineering firms and actual cost comparables. Environmental engineering firms active in wetland mitigation can be retained to prepare cost estimates on a consulting basis.

To do this, however, they must be provided sufficient information as to the type and extent of mitigation required.

Cost comparables are also a good source of replacement cost data. These costs can be discovered when researching comparable land sales that were subject to mitigation costs. If mitigation costs were involved in a potential comparable sale, those costs should be researched to derive a comparable.

As in the sales comparison approach, it is critical in the cost approach to establish like units of comparison. Replacement costs derived from engineering firms and cost comparables should reflect the same characteristics as the subject wetlands. For example, creation should be compared with creation, restoration with restoration, and enhancement with enhancement. Likewise, similar systems should be compared, i.e. Palustrine with Palustrine, Riverine with Riverine, etc. Total costs may include upland site costs if additional land purchases were required for off-site mitigation.

Once types and costs are known and equivalent units have been established, the comparison process can begin. The market value of the wetlands being appraised can then be estimated based on this cost comparison analysis.

Example

Assume you are appraising a 10-acre prime commercial site located at the corner of a busy intersection in a rapidly developing neighborhood. Your client plans to develop the site with a neighborhood shopping center to be anchored by a grocer and drug store, and this planned use is the highest and best use of the property as vacant. The site has good frontage, access, and shape requirements but has a two-acre wetland area in the middle that will be impacted by development. Overall, 80% (eight acres) of the site consists of uplands and 20% (two acres) consists of Palustrine wetlands.

The developer has obtained a permit from the necessary government agencies to develop the site as planned. The permit requires the developer to create two acres of wetlands as mitigation (compensation) for the two acres of wetlands to be developed. As such, the mitigation ratio is 1:1. The developer intends to create these wetlands in the far rear portion of the site, and the mitigation costs will be $40,000 per acre, or $80,000. In addition, the cost to fill existing wetlands to make them upland equivalents will be $10,000 per acre, or $20,000. Therefore, total mitigation costs and fill will be $100,000, or $50,000 per acre.

Fortunately, there have been several sales nearby that also required wetland mitigation for development. Research of these sales reveals buyers have been spending $40,000 to $45,000 per acre on wetland creation costs and $5,000 to $10,000 per acre for fill and compaction, and allocating a unit value of $150,000 per acre for uplands. Based on this cost and market information, the overall value of the site is estimated as follows:

$$V_o = V_u + V_w$$
$$= (\$150,000/acre \times 8 \text{ acres}) + (\$50,000/acre \times 2 \text{ acres})$$
$$= \$1,200,000 + 100,000$$
$$= \$1,300,000$$

Note that this example assumes conversations with developers reveal they placed no value on the wetlands subject to mitigation other than the cost of mitigation and fill. Therefore, no value was placed on the underlying land used in wetland creation. Such factors should always be researched.

It is important to place the value resulting from the cost approach in context by comparing it with the sales comparison approach to determine whether any entrepreneurial profit or overimprovement is present. For example, assume comparable sites have been selling for $130,000 per gross acre. Applying this unit value to the subject site area of 10 acres yields a value via the sales comparison approach (whole-to-whole method) as follows:

$$V_o = \$130,000/acre \times 10 \text{ acres}$$
$$V_o = \$1,300,000$$

In this case, the sales comparison and cost approaches yield equivalent value estimates of $1,300,000, indicating the wetland mitigation is financially feasible. However, if the sales comparison approach yields a significantly lower value estimate, then the wetland mitigation is most likely not financially feasible and would result in an overimprovement to the site. In such a case the developer would be better served by identifying a parcel more conducive to development. Therefore, the cost approach should be taken in context as compared with the sales comparison approach.

INCOME APPROACH

The income approach to market value is based on the principle of anticipation, with the underlying assumption that value is a function of anticipated future income benefits to be received from ownership. In this approach, market value is the present worth of the right to receive future income to be generated by a property.

Currently, the income approach is not generally employed in the market valuation of wetlands unless the wetlands are encumbered by leases. In these cases the income approach can be used to estimate the market value of the leased fee or leasehold interest. This most often occurs with hunting, timber, or recreation leases or submerged lands leased from the state for marina purposes. In a lease analysis, all income and expenses should be projected per contract terms and forecasted market conditions. A projection should also be made of reversionary income at future disposition (if applicable). Then the cash flows should be discounted to present value at a rate that reflects the risk associated with the lease and property involved.

Many new valuation theories based on the income approach are being developed by researchers in which various income benefits, not just those arising from contract or market rents associated with a wetland, are estimated. For example, some of the theories include the income generated by the hunting, fishing and recreation industries; the income benefits (or foregone opportunity costs) arising from the hydrological functions of wetlands; and other such income. However, as was stressed in a previous chapter, many of these benefits accrue to the public as a whole and not to the private property owner. Although the application of these new methods may be proper in estimating the *total economic value* of a wetland, it may not be proper in estimating the *market value* of a *property owner's interest* in a wetland. In the future, however, this may change as market participants and society become better informed.

SPECIAL CONSIDERATIONS

Some special considerations exist that appraisers should keep in mind when estimating the market value of an interest in wetland properties. These include the presence of timber or merchantable agriculture, endangered or threatened species of plants and animals, and restricting easements and reservations; local zoning and land use implications; and reliance on reports prepared by others.

Large wetland properties often contain merchantable timber and agriculture. For example, wetlands make excellent farmland, as evidenced by the high rate of wetlands conversion to agricultural uses, which was discussed in the introduction of this text. Although such conversion is illegal today, there are millions of acres of agriculturally altered wetlands. One case is • Silviculture Altered Wetlands (SAWs). SAWs consist of wetlands used in tree farming and typically contain merchantable timber. Technically, merchantable vegetation with an expected life of more than one year is considered a fixture and part of the real estate. Therefore, any valuation of a wetland must include the value of any such vegetation that is present. If a timber cruise or other such assessment prepared by a certified specialist in that field has not been provided for review, then the appraiser should include a special disclosure stating the lack of such information and resulting inability to estimate this value component of the real estate.

Wetlands are also often home to many species of threatened and endangered plants and animals. If such wildlife is present, then not only will the development potential of the wetland be restricted but that of adjoining uplands may also be restricted. The presence of bald eagles, snaildarters, spotted owls, and gopher turtles have precluded the use and development of large tracts of land in specific regions of the United States. Again, if a wildlife assessment prepared by a certified specialist has not been provided, then a special disclosure stating the lack of such information and the assumption that no such species are present should be clearly made by the appraiser.

It is becoming more popular for government agencies and conservation groups to purchase conservation easements and reservations in properties. This has proven to be less expensive than purchasing a fee simple title to the property while achieving the same result. Such easements and reservations typically preclude all use of a property except for specified purposes such as periodic hunting, public recreation, and some agricultural uses. If conservation easements and reservations are present, their impact on value should be considered.

With respect to zoning and land use laws, some jurisdictions base permissible development densities on a gross acreage basis while others use a usable acreage basis. This can significantly affect the highest and best use of a property. For example, assume a 20-acre site is being appraised, and 50% of the site consists of nondevelopable wetlands. The site is zoned for multifamily use and has an allowable density of 10 units per gross acre. As such, 200 units can be developed (20 acres x 10 units per acre = 200 units). However, if the zoning density is based on usable acreage, then only 100 units can be developed (20 acres x 50% usable x 10 units per acre = 100 units). This example points up the need to research the land basis of local zoning and land use laws.

Guide Note 6 of the Appraisal Institute's Standards of Professional Appraisal Practice provides insight into the use and reliance of reports prepared by others. In general, this guide note classifies reports within four categories: General Informational Reports, Reports Prepared by Licensed or Certified Non Real Estate Appraisal Professionals, Reports

Prepared by Other Non Real Estate Appraisal Professionals, and Other Reports. The level of due diligence for appraisers varies depending on who prepared the report.

With regard to wetland valuation, appraisers generally rely on wetland surveys, soil tests, hydrology studies, plant and animal inventories, timber cruises, and other such reports. To ensure the competence of those who prepare these reports, and thus the reliability of the information within them, appraisers should accept only reports prepared by licensed professionals in these particular fields. In addition, a disclosure of reliance on these reports and a statement acknowledging a lack of knowledge and/or experience in these areas must be made in any appraisal.

SUMMARY

The three valuation approaches may be applied in the appraisal of wetlands. In the sales comparison approach, three methods of comparison analysis can be applied: the whole-to-whole method, the sum-of-the-parts method, and a residual method. It is advisable to use at least two of these methods to provide extra support to the market valuation process. The methods applied should be determined by the data available in the market. The key to all three methods is to identify like units of comparison.

In the cost approach, cost is the unit of comparison. The cost approach has become applicable in the appraisal of wetlands on prime properties subject to mitigation for development. This application is a result of government intervention in the market and recent scientific discoveries. The key to this approach, as in all the market valuation approaches, is to establish like units of comparison and apply sound judgment. This approach is generally not applicable when appraising sizable tracts.

Lastly, the income approach may be applicable if the property is encumbered by a lease.

When applying these approaches, an appraiser should keep in mind some special considerations. These include the presence of merchantable agricultural fixtures; the presence of endangered or threatened species of plants and animals; restricting conservation easements or reservations; local zoning and land use laws; and the reliance on reports prepared by others. Such considerations may have a significant impact on value.

Chapter 5

Future Trends:
Wetland Mitigation Banking

Given the ever changing legal and scientific landscape, the market valuation of wetlands poses an ongoing challenge to real estate professionals. One new development that will likely have a huge impact on wetland valuation is the emergence of wetland mitigation banking.

Developed in the early 1980s by the U.S. Fish and Wildlife Service, wetland mitigation banking is based on a similar concept used to offset air pollution impacts. The concept of wetland mitigation banks emerged in response to shortcomings of previous wetland mitigation efforts. Government agencies discovered numerous failures among wetlands that had been created or restored in mitigation proceedings. In fact, studies showed that some mitigation projects were never even performed. The primary reason for these failures was not the poor design or construction of wetlands but inadequate maintenance and monitoring practices.

To correct this problem, mitigation banks are being established as an alternate means of mitigating impacts. Such banks will generally consist of large areas of created or restored wetlands. When proposed development impacts a wetland and off-site mitigation is required, then rights in the bank can be purchased. These rights are either acreage-based or value-based. The bank of created or restored wetlands can be used to offset the loss of wetlands resulting from development. In this way, the goal of no net loss of wetlands can be achieved.

The formation of wetland mitigation banks requires special permits, and such banks must be owned and managed by competent and adequately funded parties. Theoretically, the owners will have the expertise and funds to make a long-term conservation effort successful.

Banks can be constructed for the sole use of one party (self financed); they can be constructed by a group who then sell wetland credits to the general public (market or fee based); or they can be in the form of a trust fund that developers pay into and which will be used to carry out compensation activities at a later date. . . . Since 1980, approximately 45 wetland mitigation banks have been created by government agencies and private organizations, with many more in various stages of planning or development.[1]

Banks can be privately or publicly owned, and credits may be used for internal purposes or sold in the market to others. An example of a public bank is found in Long Beach, California, where the municipality's port authority has established two mitigation banks, one of 29 acres in the Upper Newport Bay Ecological Reserve to mitigate adverse impacts from the development of a pier in Newport Bay and a second in the Seal Beach National Wildlife Refuge to mitigate impacts from a pier landfill in Anaheim Bay.[2]

The Walker Ranch bank, located in Florida and developed by the Disney Development Corporation, is an example of a self-financed private bank. The bank serves to compensate for wetland impacts from continuing development at Walt Disney World. The bank is managed by the Nature Conservancy.[3]

A private, entrepreneurial bank is being developed in northeast Florida by a group of investors who are applying for a wetland mitigation bank permit. The investors have optioned a 380-acre dairy farm in the area that consists of a converted wetland. If the permit is issued, the developers plan to remove hydrological barriers and restore the farm to its natural wetland state. They then plan to sell acreage-based mitigation credits from the wetland bank to local developers and other parties. In this case, the bank will be a for-profit operation.

As more wetland mitigation banks are established and sell mitigation rights, the interaction of buyers and sellers will establish the value of such

---

1. Steven Crookshank, *Air Emissions Banking and Trading: Analysis and Implications for Wetland Mitigation Banking,* Research Study #074 (American Petroleum Institute, February 1994), p. 1.

2. Robert Anderson and Mark Rockel, *Economic Valuation of Wetlands,* Discussion Paper #065 (American Petroleum Institute, April 1991).

3. P.L. Wilkey, R.C. Sundrell, and D.C. Hayes, *Wetland Mitigation Banking for the Oil and Gas Industry: Assessment, Conclusions, and Recommendations* (Washington, D.C.: U.S. Department of Energy, January 1994), p. 41.

rights in the market as well as the value of the banks themselves. Over time, wetland mitigation banks are expected to have a significant effect on the wetlands marketplace by promoting a more homogeneous mitigation product, better information, and greater efficiency. In the future, it is likely appraisers will be called upon to estimate the market value of a wetland mitigation bank or value wetlands on a site subject to mitigation as a function of wetland bank credits. Such assignments will require special study and expertise.

SUMMARY

While the concept of wetland banking is new, its use as a form of mitigation is attractive to both developers and government. Developers like the concept because the banks will free them from a long-term commitment in wetland maintenance and shorten the time associated with mitigation proceedings. Government agencies like the concept because the banks will protect huge blocks of wetland ecosystems and ensure competent long-term management practices. Therefore, both sides of the aisle appear to be excited about the future of wetland mitigation banking.

Not only is the market value of wetlands dynamic, but so is the wetland market in general. Since wetlands are such a highly regulated type of real estate, it is likely that changes in the market will continue. As such, it is important to recognize current valuation methods and continue to build upon them with new discoveries to best solve the market valuation problem.

Glossary

acquifer. A water-bearing stratum of permeable rock, sand, or gravel.

anaerobic. Lacking oxygen.

cost approach. A set of procedures through which a value indication is derived for the fee simple interest in a property by estimating the current cost to construct a reproduction of, or replacement for, the existing improvements.

estuarine wetland system. Deepwater tidal habitats and adjacent tidal wetlands that are usually semi-enclosed by land but have open, partly obstructed, or sporadic access to the open ocean, and in which ocean water is at least occasionally diluted by freshwater runoff from the land. The salinity may be periodically increased above that of the open ocean by evaporation. Along some low-energy coastlines there is appreciable dilution of sea water. Offshore areas with typical estuarine plants and animals, such as red mangroves and eastern oysters, are also included in the Estuarine system.

executive order. A rule or regulation issued by the President of the United States which has the force of law.

fee simple estate. Absolute ownership unencumbered by any other interest or estate, subject only to the limitations imposed by the governmental powers of taxation, eminent domain, police power, and escheat.

hydric soils. Soils which are saturated, flooded, or ponded long enough during the growing season to develop anaerobic conditions in the upper part.

hydrology. The science dealing with the properties, distribution, and circulation of water on the surface of the land, in the soil and underlying rocks, and in the atmosphere.

hydrophytic vegetation. Vegetation that can grow despite anaerobic soil conditions and periodic water inundation.

income capitalization approach. A set of procedures through which an appraiser derives a value indication for an income-producing property by converting its anticipated benefits (cash flow and reversion) into property value.

lacustrine wetland system. Wetlands and deepwater habitats with all of the following characteristics: 1) situated in a topographic depression or a dammed river channel; 2) lacking trees, shrubs, persistent emergents, emergent mosses or lichens with greater than 30% aerial coverage; and 3) total area exceeds eight hectares (20 acres). Similar wetland and deepwater habitats totaling less than eight hectares are also included in the Lacustrine system if an active wave-formed or bedrock shoreline feature makes up all or part of the boundary, or if the water depth in the deepest part of the basin exceeds two meters (6.6 feet) at low water.

leased fee estate. The interest held by a lessee (the tenant) through a lease conveying the right of use and occupancy for a stated term under certain conditions.

marine wetland system. The open ocean overlying the continental shelf and its associated high-energy coastline. Marine habitats are exposed to the waves and currents of the open ocean and the water regimes are determined primarily by the ebb and flow of ocean tides. Shallow coastal indentations or bays without appreciable freshwater inflow and coasts with exposed rocky islands that provide the mainland with little or no shelter from wind and rain are also considered part of the Marine system because they generally support typical marine biota.

market value. The most probable price which a property should bring in a competitive and open market under all conditions requisite to a fair sale, the buyer and seller each acting prudently and knowledgeably, and assuming the price is not affected by undue stimulus. Implicit in this definition is the consummation of a sale as of a specified date and the passing of title from seller to buyer under conditions whereby:

1. buyer and seller are typically motivated;
2. both parties are well informed or well advised, and acting in what they consider their own best interests;
3. a reasonable time is allowed for exposure in the open market;
4. payment is made in terms of cash in U.S. dollars or in terms of financial arrangements comparable thereto; and
5. the price represents the normal consideration for the property sold unaffected by special or creative financing or sale concessions granted by anyone associated with the sale.

Palustrine wetland system. All nontidal wetlands dominated by trees, shrubs, persistent emergents, emergent mosses or lichens. It also includes wetlands lacking such vegetation, but with all of the following four characteristics: 1) area less than eight hectares (20 acres); 2) active wave-formed or bedrock shoreline features lacking; 3) water depth in the deepest part of the basin less than two meters at low water; and 4) salinity due to ocean-derived salts less than 0.5 parts per thousand.

replacement cost. The estimated cost to construct, at current prices as of the effective appraisal date, a building [wetland] with utility equivalent to the building [wetland] being appraised, using modern materials and current standards, design, and layout.

reproduction cost. The estimated cost to construct, at current prices as of the effective date of appraisal, an exact duplicate or replica of the building [wetland] being appraised, using the same materials, construction standards, design, layout, and quality of workmanship and embodying all the deficiencies, superadequacies, and obsolescence of the subject building [wetland].

riverine wetland system. All wetlands and deepwater habitats contained within a channel, with two exceptions: 1) wetlands dominated by trees shrubs, persistent emergents, emergent mosses, or lichens, and 2) habitats with water containing salts in excess of 0.5 parts per thousand. A channel is an open conduit either naturally or artificially created which periodically or continuously contains moving water, or which forms a connecting link between two bodies of standing water.

sales comparison approach. A set of procedures in which a value indication is derived by comparing the property being appraised to similar properties that have been sold recently, applying appropriate units of comparison, and making adjustments to the sale prices of the comparables based on the elements of comparison.

swampbuster provision. A provision in the Food Security Act to discourage the agricultural conversion of wetlands; it made any farmer who converts wetlands to agricultural uses ineligible for price supports, payments, crop insurance, or government-insured loans.

USPAP. Uniform Standards of Professional Appraisal Practice as established by The Appraisal Foundation.

wetland. Land that has a predominance of hydric soils and that is inundated or saturated by surface or groundwater at a frequency and duration sufficient to support, and that under normal conditions does support, a prevalence of hydrophytic vegetation typically adapted for life in saturated soil conditions.